WEIRD ANIMAL SCIENCE

ANIMAL COMMUNICATION

THE BREAKDOWN

VIRGINIA LOH-HAGAN

45TH PARALLEL PRESS

Published in the United States of America by Cherry Lake Publishing Group
Ann Arbor, Michigan
www.cherrylakepublishing.com

Photo Credits: © Adisak Mitrprayoon/iStock, cover, 1; © Polina Tomtosova/Shutterstock, cover, 1; © Rama Aadhithan/Shutterstock, 6; © chris276644/Shutterstock, 9; © Yury Nikolaev/Shutterstock, 10; © Leonne Sa Fortes/Shutterstock, 12; © Santi Rodriguez/Shutterstock, 14; © Andrew Shevchuk/Shutterstock, 15; © Light-Dew/Shutterstock, 16; © SuperMoo Varavut/Shutterstock, 17; © Richard Susanto/Shutterstock, 18; © Abuntcreative/Shutterstock, 19; © Image courtesy of The Alex Foundation, 21

Graphic Element Credits: Graphic Element Credits: Cover, multiple interior pages: © WindAwake/Shutterstock; © Theo_hrm/Shutterstock; © Rodin Anton/Shutterstock; © Petr Vaclavek/Shutterstock; © Nikolaeva/Shutterstock

45th Parallel Press is an imprint of Cherry Lake Publishing Group.

Library of Congress Cataloging-in-Publication Data has been filed and is available at catalog.loc.gov

Cherry Lake Publishing Group would like to acknowledge the work of the Partnership for 21st Century Learning, a Network of Battelle for Kids. Please visite Battelle for Kids online for more information.

Printed in the United States of America

Dr. Virginia Loh-Hagan is an author, university professor, and former classroom teacher. She's currently the Director of the Asian Pacific Islander Desi American Resource Center at San Diego State University. She loves science-fiction stories. She lives in San Diego with her very tall husband and very naughty dogs.

TABLE OF CONTENTS

BREAK IT DOWN

1. DECODE IT

- Look at each letter.
- Look at letter patterns.
- Blend sounds.
- Try other possible sounds.

2. BUILD VOCABULARY

- Listen to the word.
- Match the sounds to a word you know.
- Look for context clues for words you don't know.
- Look up the meaning in a glossary or dictionary.

3. MAKE CONNECTIONS

- Name other examples.
- Draw the idea.
- Compare ideas.
- Ask questions.

TEST IT OUT

WORDS YOU'LL LEARN

Look over the list of vocabulary words. Are there any you already know?

avian
coordinate
pheromones
vocal cords

GET STARTED

The letter *r* can change the sound of a vowel that comes before it. Keep the vowel and letter *r* together when you read words.

- Look at the letters.
- Chunk long words to read them.
- Try different sounds for *r*-controlled vowels.
- For example, the letters *er* in longer words can say /er/, /air/, or /ear/.

The word *cords* has an *r*-controlled *o*.
It is pronounced: KORDZ.

Animals communicate for different reasons.

They find mates. They defend their space.

They **coordinate** group action. Coordinate means set up.

They care for their young. They show power.

DECODE IT

The word *coordinate* has 4 vowel sounds.

- Break it into chunks: co-or-di-nate.
- The first vowel is a long *o*.
- The next is an *r*-controlled *o* and says "or."
- The next vowel sound says "uh."
- The last vowel is a long *a*.
- *Coordinate* is pronounced koh-OR-duh-nayt.

Animals communicate in different ways.

Some make sounds. Some use body language.

MAKE CONNECTIONS

Think about how animals use body language to communicate.

- How do people use body language to communicate?
- What can it mean if a person crosses their arms? What if they raise their eyebrows?

Some send signals. Some use touch.

Some send **pheromones**.

Pheromones are animal scents.

Animals release their pheromones through sweat and pee.

DECODE IT

The word *pheromones* has 3 vowel sounds.

- Break it into chunks: pher-o-mones.
- The *ph* says "fff."
- The first vowel sound is an *r*-controlled *e* that says "air."
- The next vowel sound says "uh."
- The last vowel is a long *o*.
- *Pheromones* is pronounced FAIR-uh-mohnz.

Other animals in the same species smell the pheromones.

This triggers a response.

Ants and bees use pheromones, too.

They use them to mark food trails.

Many animals, including elephants, can make human sounds.

But parrots are the best at this.

Parrots copy human and animal sounds.

They have a special voice box.

Humans push air over **vocal cords** to create sounds. Vocal cords are tissue flaps.

But parrots use their throat muscles.

They direct air over their windpipes.

They change the depth and shape of their windpipes.

Alex was an African grey parrot.

An animal scientist studied Alex for 30 years.

His name stands for **Avian** Language Experiment.

Avian means bird.

BUILD VOCABULARY

Avian (AY-vee-uhn) means related to birds.

- This comes from Latin *avis*, which means bird.
- Avian dinosaurs are bird-like dinosaurs.
- Other forms: aviator, aviation

Alex was trained.

He knew more than 100 words.

PUT IT TOGETHER

Detail:
Some animals use body language to communicate.

Detail:
Parrots can even learn human words.

Detail:
Some animals use pheromones to communicate.

Main Idea:
Animals communicate in different ways for different reasons.

Detail:
Some animals can make human sounds.

Detail:
Some animals use sounds to communicate.

LEARN MORE

BOOKS

Jenkins, Steve. *The Animals Book*. Boston, MA: HMH Books for Young Readers, 2013.

Lerwill, Ben, and Sarah Walsh (illus.). *WildLives: 50 Extraordinary Animals That Made History*. New York, NY: Atheneum Books for Young Readers, 2020.

Watson, Galadriel, and Samantha Dixon (illus.) *Running Wild: Awesome Animals in Motion*. Toronto, ON: Annick Press, 2020.

ONLINE

Search these online sources with an adult:

- National Geographic Kids | Animals
- Britannica Kids | Animal Kingdom

GLOSSARY

avian (AY-vee-uhn) relating to birds

coordinate (koh-OR-duh-nayt) to set up common action

pheromones (FAIR-uh-mohnz) scents given off by some animals that cause a social response in members of the same species

vocal cords (VOH-kuhl KORDZ) tissue folds in the human throat that vibrate and produce sounds used in speech

INDEX